Mathematics WARM-UPS

Grade 6

WALCH EDUCATION

1 2 3 4 5 6 7 8 9 10
ISBN 978-0-8251-7147-5

J. Weston Walch, Publisher
Portland, ME 04103
www.walch.com
Printed in the United States of America

Table of Contents

Introduction

Mathematics Warm-Ups for Common Core State Standards, Grade 6 is organized into five sections, composed of the domains for Grade 6 as designated by the Common Core State Standards Initiative. Each warm-up addresses at least one of the standards within the following domains.

- Ratios and Proportional Relationships
- The Number System
- Expressions and Equations
- Geometry
- Statistics and Probability

The Common Core Mathematical Practices standards are another focus of the warm-ups. All of the problems require students to "make sense of problems and persevere in solving them," "reason abstractly and quantitatively," and "attend to precision." Students must "look for and make use of structure" when finding lowest common denominators and greatest common multiples. Students have opportunities to "use appropriate tools strategically" when they use 10×10 grids to examine proportional relationships or graph paper to explore area and equivalence. A full description of these standards can be found at http://www.walch.com/CCSS/00006.

The warm-ups are organized by domains rather than by level of difficulty. Use your judgment to select appropriate problems for your curriculum.* The problems are not necessarily meant to be completed in consecutive order—some are stand-alone, some can launch a topic, some can be used as journal prompts, and some refresh students' skills and concepts. All are meant to enhance and complement your Grade 6 mathematics program. They do so by providing resources for those short, 5- to 15-minute interims when class time might otherwise go unused.

* You may select warm-ups based on particular standards using the Standards Correlations table.

About the CD-ROM

Mathematics Warm-Ups for Common Core State Standards, Grade 6 is provided in two convenient formats: an easy-to-use reproducible book and a ready-to-print PDF on a companion CD-ROM. You can photocopy or print activities as needed, or project them on a screen via your computer.

The depth and breadth of the collection give you the opportunity to choose specific skills and concepts that correspond to your curriculum and instruction. Use the table of contents and the standards correlations to help you select appropriate tasks.

Suggestions for use:

- Choose an activity to project or print out and assign.
- Select a series of activities. Print the selection to create practice packets for learners who need help with specific skills or concepts.

Standards Correlations

Mathematics Warm-Ups for Common Core State Standards, Grade 6 is correlated to five domains of CCSS Grade 6 mathematics. The page numbers, titles, and standard numbers are included in the table that follows. The full text of the CCSS mathematics standards for Grade 6 can be found in the Common Core State Standards PDF at http://www.walch.com/CCSS/00001.

Page number	Title	CCSS addressed
Ratios and Proportional Relationships		
1	Jumping Jellies!	6.RP.1
2	Solving Proportions	6.RP.2
3	Money, Money, Money	6.RP.3a
4	Birthday Roses	6.RP.3b
5	Paolo's Pizza Pricing	6.RP.3b
6	Stamping Around	6.RP.3b
7	Faster Than a Speeding Bullet	6.RP.3b
8	Let It Snow	6.RP.3b
9	Square Pizza	6.RP.3c
10	Playing with Proportions	6.RP.3c
11	How Sweet It Is	6.RP.3d
12	Oil and Vinegar	6.RP.3d
13	Find the Distance	6.RP.3d
The Number System		
14	Ribbon and Bows	6.NS.1
15	Baking Blueberry Pies	6.NS.1
16	Super Sub Sandwich	6.NS.1
17	Fun with Fractions	6.NS.1

(*continued*)

Page number	Title	CCSS addressed
18	Using 10 × 10 Grids	6.NS.3
19	What Day of the Week?	6.NS.4
20	Greatest Common Factor	6.NS.4
21	Library Day	6.NS.4
22	Cicada Cycles	6.NS.4
23	Counting Cookies	6.NS.4
24	Factors and Multiples	6.NS.4
25	Forward and Back	6.NS.5
26	Balloon Flight	6.NS.5
27	Scuba Duba Doo	6.NS.6c
Expressions and Equations		
28	As Easy As ABC	6.EE.1
29	Powers and Exponents	6.EE.1
30	A Number Puzzle	6.EE.2a, 6.EE.3
31	Do You Speak Math?	6.EE.2a
32	Writing Equations	6.EE.2a
33	Collette's Chocolates	6.EE.2c
34	Evaluating Simple Expressions I	6.EE.2c
35	Evaluating Simple Expressions II	6.EE.2c
36	Using Properties	6.EE.3
37	Border Tiles	6.EE.3, 6.EE.4
38	Solving Equations	6.EE.5
39	Inequalities I	6.EE.5
40	Dividing by Fractions	6.EE.6
41	Inequalities II	6.EE.8

(*continued*)

Jumping Jellies!

Use the information from the scenario below to answer the questions.

Annie has a big bowl of jelly beans. She has the following number of each flavor:

25 strawberry

25 lemon

50 pineapple

25 blueberry

75 lime

1. How many strawberry jelly beans are there?

2. How many blueberry jelly beans are there?

3. How does the number of strawberry jelly beans compare to the number of blueberry jelly beans?

4. How many pineapple jelly beans are there?

5. How many jelly beans are there altogether?

6. How does the number of pineapple jelly beans compare to the total number of jelly beans? What fraction of the jelly beans are pineapple?

Solving Proportions

Remember that a proportion is two equal ratios.

$$\frac{a}{b} = \frac{c}{d}$$

To solve a proportion, cross-multiply.

$$\frac{a}{b} \times \frac{c}{d}$$

$$ad = bc$$

Solve the following proportions for x.

1. $\frac{x}{3} = \frac{6}{9}$

2. $\frac{1}{2} = \frac{x}{5}$

3. $\frac{3}{4} = \frac{8}{x}$

4. $\frac{x-3}{4} = \frac{x}{8}$

Money, Money, Money

You got a great job babysitting for $8 an hour. Fill out the table below to figure out how much you would make if you worked a full 8-hour day.

Number of hours	Amount earned
1	$8
2	
3	$24
	$32
5	
6	
	$56
8	

Birthday Roses

Read the scenario and answer the question that follows.

> Alonzo is planning to purchase roses for his mother on her birthday. He has seen them advertised at 12 roses for $15.00 and 20 roses for $23.00.

Which is the better buy? Show your work and explain your reasoning.

Paolo's Pizza Pricing

Paolo has just started working at his Uncle Antonio's pizza parlor. He is trying to figure out which size meat lover's pizza provides the best value.

Meat Lover's Special

Size	Price
9-inch round pizza	$10.50
12-inch round pizza	$15.00
18-inch round pizza	$19.00

Which pizza listed on the menu provides the best value? Write a few sentences that explain your reasoning.

Stamping Around

In 2012, a book of 20 first class stamps cost $9. How much would one stamp cost? Show your work and write your answer below.

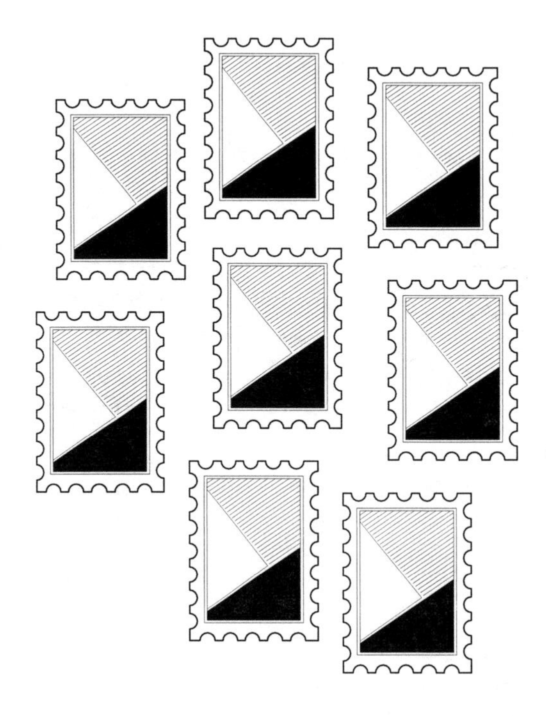

Faster Than a Speeding Bullet

Anna's family took a trip to the beach for a vacation. Anna's mom drove for 4 hours at a constant speed and went 240 miles. How fast did Anna's mom drive? (Distance = Rate × Time) Show your work and write your answer below.

Let It Snow

Read the following scenario. Use the information in it to answer the question.

> During the winter months, Esteban makes $25 for every snowy walkway he shovels. So far this winter, he has earned $250 from shoveling. How many walkways has he shoveled?

Write your answer and show your work below.

Square Pizza

Lin, Lon, Lu, and Lau have ordered a pizza from the Tip Top Pizza Palace. The pizzas only come in one size and are in the shape of a square. Lau has just had a blueberry smoothie and is not very hungry, but she thinks that she might eat 10% of the pizza. Lon is famished and thinks that he might eat half of the pizza. Lin thinks she might eat about 35% of the pizza, and Lu thinks he might eat 15%. If the four friends eat the portions that they have predicted, what percent of the pizza will remain? Justify your thinking using a 10×10 grid like the one below.

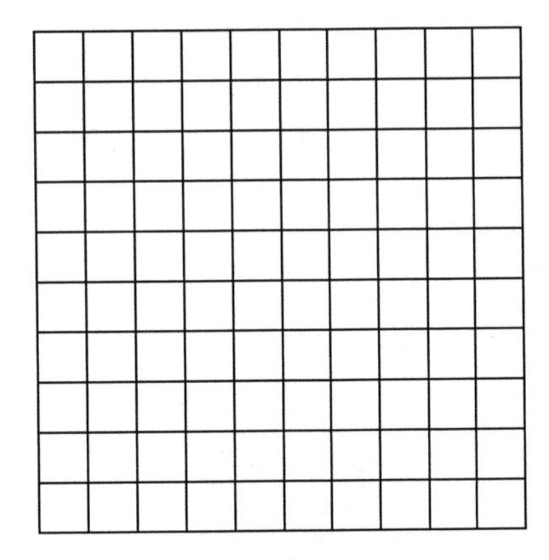

Playing with Proportions

Jamal got his math test back and saw his teacher had written $\frac{16}{20}$ on the test. Jamal's friend said, "Wow, Jamal, you got 80%." Is Jamal's friend correct? Check to see if $\frac{16}{20}$ and $\frac{80}{100}$ are equivalent fractions. If so, then Jamal did get 80%. Show your work below.

How Sweet It Is

Use what you know about fractions and percents to solve.

There were six groups of students in science class. Each group was given a bag with sugar and sand in it. The groups were asked to figure out the ratio of sugar in the bags they were given. Here are the groups' answers:

$30\%, 23\%, \frac{1}{5}, 0.18, \frac{2}{9}, 0.25$

Put the numbers in order from least to greatest.

NAME:

Oil and Vinegar

Read the scenario. Use the information in it to complete the problem.

> You are making a salad dressing that calls for a ratio of 3 parts oil to 2 parts vinegar. If you are using 12 tablespoons of oil, how much vinegar will you need?

Write your answer and show your work below. Make a drawing to illustrate your solution.

Find the Distance

Use what you know about converting measurements to solve.

Maya is measuring her bedroom, but her tape measure only shows lengths in feet. She measures and finds that the room is 15 feet long. How many yards long is Maya's room?

Make a sketch to show your thinking.

Ribbon and Bows

Veronica is making decorative bows for a craft project. She has 7 yards of velvet ribbon. Each bow requires $\frac{3}{4}$ of a yard of ribbon. How many bows will she be able to make with the ribbon she has? Will she have any ribbon left? Show your work in the space below and explain your thinking.

NAME:

THE NUMBER SYSTEM
CCSS 6.NS.1

Baking Blueberry Pies

Mrs. Berry is famous for her pies. She has been making pies in her bakery for many years. She knows that it takes $1\frac{2}{3}$ cups of flour to make her special pie crust. She buys flour in 25-pound bags and knows that each pound contains about 3 cups of flour. How many pies can she expect to make from a 25-pound bag of flour?

Super Sub Sandwich

The students at Abigail Adams Middle School tried to make the biggest sub sandwich on record. They made a sandwich that was $12\frac{3}{4}$ feet long. After they were told that this did not break the record, they decided to divide the sandwich into smaller portions and share it with other students.

1. If each portion was $\frac{1}{2}$ foot long, how many students would get a portion?

2. If each portion was $\frac{3}{4}$ foot long, how many students would get a portion?

3. For each problem above, explain and illustrate your thinking using a diagram.

Fun with Fractions

Answer the following word problem.

Greg bought $2\frac{1}{2}$ pounds of roast beef. He is making sandwiches with $\frac{1}{4}$ pound of roast beef in each sandwich. How many sandwiches can Greg make? Will he have any roast beef left over? Draw a picture to show your thinking.

Using 10 × 10 Grids

Micaela is learning about the connection between decimals and fractions. She has made the following diagram in her notes to represent multiplying decimals less than 1. What multiplication sentence(s) might be represented by the darker shaded area of her drawing?

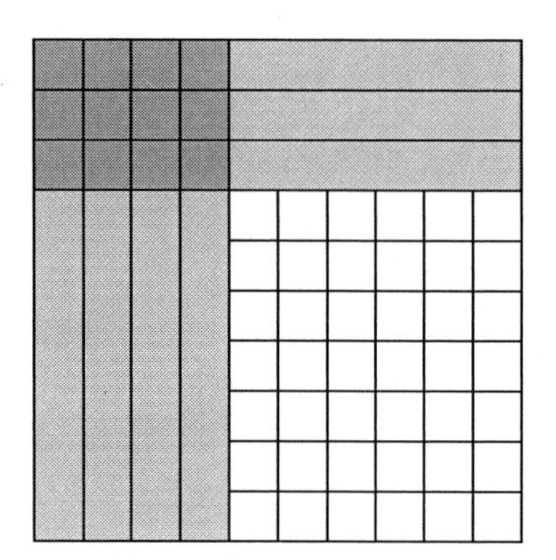

What Day of the Week?

Saari and Anatole are cooped up inside because of bad weather. Their mother, a middle-school teacher, poses a puzzler for them. She says that Anatole was born 759 days after Saari, with no leap year between the two birthdays. If Saari was born on a Wednesday, on what day of the week was Anatole born? Find the day of the week and explain your strategy for finding it.

Greatest Common Factor

Find the greatest common factor (GCF) of the numbers that follow. Show your work in the space provided below.

360

336

1,260

Describe the strategy you used to find the GCF. Will your strategy work for all situations? Explain.

Library Day

Marika, Bobbie, and Claude all visit their local public library on a regular schedule. Marika visits every 15 days, Bobbie goes every 12 days, and Claude goes every 25 days. If they are all at the library today, in how many days from now will they all be there again? What strategy did you use to find the number of days? What does your answer represent?

Cicada Cycles

Cicadas are insects that eat plants. Stephan's grandfather told him that some cicadas have 13-year or 17-year cycles. He said that one year, the cicadas were so numerous on his family farm that they ate all the crops. Stephan guessed that perhaps both the 13-year and the 17-year cicadas came out that year.

1. If we assume Stephan is correct, how many years will have passed when the 13-year and 17-year cicadas come out together again?

2. Imagine that there are 12-year, 14-year, and 16-year cicadas, and they all come out this year. How many years will pass before they all come out together again?

3. Explain how you got your answer. What number does your answer represent?

Counting Cookies

Read the word problem and answer the questions. Remember to show your work and explain your answers.

Hannah baked 120 peanut butter chocolate chip cookies. She wants to put the same number of cookies in each bag.

1. Could she put 10 cookies in each bag?

2. What about 5 cookies?

3. 9 cookies?

4. 3 cookies?

5. 20 cookies?

6. 7 cookies?

Factors and Multiples

Part 1

List all the numbers that divide evenly into 12 from least to greatest.

List all the numbers that divide evenly into 18 from least to greatest.

The numbers that divide evenly into a number with no remainder are called its factors.

Circle the factors that 12 and 18 have in common.

12: 1, 2, 3, 4, 6, 12

18: 1, 2, 3, 6, 9, 18

Part 2

List the numbers you get when you multiply 8 by 1, 2, 3...all the way up to 10.

List the numbers you get when you multiply 6 by 1, 2, 3...all the way up to 10.

The numbers that you get when you multiply a number by any whole number are its multiples.

Underline the multiples they have in common.

8: 8, 16, 24, 32, 40, 48, 56, 64, 72, 80

6: 6, 12, 18, 24, 30, 36, 42, 48, 54, 60

NAME:

Forward and Back

Read through the following situation with a partner. Then answer the questions.

You and a friend are both standing on the playground. You walk forward 7 paces. Your friend walks backward 7 paces. How would you compare the distances you each covered? Are they the same? Different? Explain.

Balloon Flight

Read the scenario and use the information to answer the question.

> A helium balloon was released in San Francisco, California, which is at sea level. The balloon rose 10,000 feet and was carried eastward by the winds. Then it lost 4,720 feet of altitude and came down in Denver, Colorado. What is Denver's altitude in feet?

Solve the problem and show your work in the space below.

Scuba Duba Doo

Read the scenario that follows. Use the information in it to complete the problem by drawing a picture.

Tanya was standing on top of a mountain that was 2,400 feet high. Olga was scuba diving 30 feet below sea level. Draw a picture to show how many feet apart Tanya and Olga are.

As Easy As ABC

By using different operations and exponents, you can get a variety of answers with the same numbers. For example, let's say you have $a = 4$, $b = 5$, and $c = 2$.

Here are some examples:

$abc = (4)(5)(2) = 40$

$(ab)^c = (4 \bullet 5)^2 = 20^2 = 400$

$a \div b - c = 4 \div 3 = \frac{4}{3} \text{ or } 1\frac{1}{3}$

Come up with three different answers using all three letters if $a = 6$, $b = 2$, and $c = 3$. (Fractions are okay.) Show your work below.

Remember the order of operations when you write and evaluate these expressions.

1. Simplify inside parentheses.
2. Simplify expressions with exponents.
3. Do multiplication and division from left to right.
4. Do addition and subtraction from left to right.

Powers and Exponents

A number such as 3^2 has a base number (3) and an exponent (^{2}). A number that is shown as a combination of a base and an exponent is called a power. Powers are a shorter way of writing some longer expressions.

Examples:

$5 \times 5 = 5^2$

$6 \times 6 \times 6 \times 6 = 6^4$

$a \bullet a \bullet a = a^3$

$x \bullet x \bullet x \bullet y \bullet y \bullet y \bullet y = x^3y^4$

Show the following in exponent form.

1. $9 \times 9 \times 9 \times 9 =$

2. $h \bullet h \bullet h \bullet h \bullet h \bullet h \bullet h =$

3. $m \bullet m \bullet n \bullet n \bullet n \bullet n \bullet n =$

4. $10 \times 10 \times 10 \times 10 =$

Answer the following.

5. What is the value of the expression 5^4?

6. What is a^3 if $a = 4$?

A Number Puzzle

Follow the steps below to complete this number puzzle.

> Choose any number. Add the number that is 1 more than your original number. Add 11. Divide by 2. Subtract your original number. What is your answer?

Do the puzzle again for other numbers. Why do you get the answers that you get? Will this always work? Show your work and answer the questions in the space below.

Do You Speak Math?

Mathematics can be thought of as another language. Really, it's true! Sometimes you have to "translate" from math to words, and sometimes you have to "translate" from words to math.

Fill in the following chart by "translating" from one column to the other. Some of the rows are filled in for you.

Math	Words
$y-2$	y minus 2
	7 plus z
$2p$	
	the product of m and 3
	the quotient of c and 5
$\frac{x}{4}$	
$t+9$	
	8 more than v
$6+b$	6 increased by b
$10-r$	

Writing Equations

An equation is a mathematical statement that contains an equal sign (=). An algebraic equation is a mathematical statement that contains the equal sign as well as algebraic expressions. For example, the statement $x + 1 = 5$ is an algebraic equation.

Write an algebraic equation for each of the following.

1. Your height is Billy's height plus 13 inches.

2. The total cost of a meal is $3 per hamburger plus $2 per milkshake.

3. The total length of a road race is half the distance of your last race minus 500 meters.

4. The total cost for a group of people to go to the movies is $10.50 per person.

Collette's Chocolates

Collette works as the marketing manager for a small chocolate-making company. She has data that shows that the company's orange creams and French mints are the most recent best sellers. The French mints are a little more popular than the orange creams. Her supervisor has asked her to create new packaging for a Valentine's Day sales promotion. Collette has come up with the following plan for different-sized boxes of chocolates. The round shapes represent orange creams. The squares represent French mints.

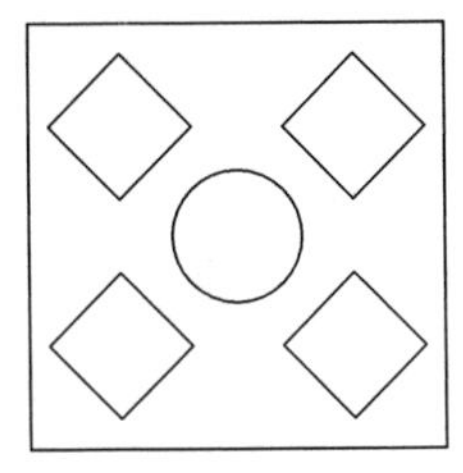

2×2 box

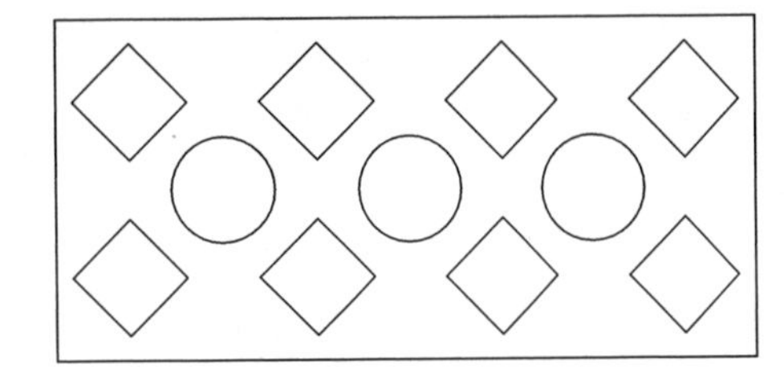

2×4 box

1. If Collette continues her pattern and she increases both the number of rows and/or the number of columns, what would a 4×6 box of chocolates look like? Draw several more boxes of chocolates that you think will continue Colette's design.

2. How many orange creams would be in the 4×6 box? Write a rule for finding the number of orange creams if you know the rows and columns of French mints.

Evaluating Simple Expressions I

Evaluate each expression below for the given value of x.

1. $4.5x - 12$ for $x = 2$

2. $-5 - x$ for $x = \frac{1}{2}$

3. $2x^2$ for $x = 8$

4. $x^2 + 2.5$ for $x = 1.5$

5. $(x - 2.5)(x + 35)$ for $x = 2.5$

6. $x(27 - x)$ for $x = 3$

7. $\frac{49}{x^2}$ for $x = -7$

8. $5x^2 + 3x - 10$ for $x = 2$

9. $\frac{x}{4}$ for $x = 8$

10. $0.5x^2 + 2x - 20$ for $x = 10$

Evaluating Simple Expressions II

Evaluate each expression below for the given value of *x*.

1. $42 - 3x$ for $x = 6$

2. $13 - 3x^2$ for $x = 1$

3. $4x^2 + 13$ for $x = -10$

4. $6x^2 + x - 2$ for $x = 2$

5. $(x - 2.5)(x + 5)$ for $x = 0$

6. $x(27 - x)$ for $x = -3$

7. $\frac{3(15 - x)}{2x}$ for $x = 10$

8. $8x - 3x(6 - x)$ for $x = 0$

9. $\frac{x^2}{4}(x + 8)$ for $x = -8$

10. $-2x^2 + 5x + 12$ for $x = -4$

Using Properties

The expressions below have been rewritten using several of the basic number properties.

Distributive Property

$4(x + 3) = 4x + 12$

Commutative Property of Multiplication

$19 \cdot 4 = 4 \cdot 19$

Associative Property of Addition

$8 + (2 + 39) = (8 + 2) + 39$

Rewrite the expressions below using the number properties illustrated above.

1. $(4 + 91) + 9 =$

2. $3(10 + 2) =$

3. $84 \cdot 2 =$

NAME:

EXPRESSIONS AND EQUATIONS
CCSS 6.EE.3, 6.EE.4

Border Tiles

Landscapers often use square tiles as borders for garden plots and pools. The drawing below represents a square pool for goldfish surrounded by 1-foot square tiles.

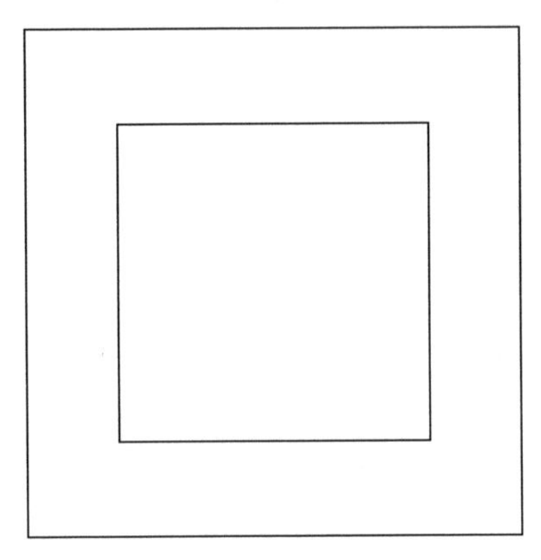

Maria thinks that the total number of tiles around the outside of the pool could be represented by $(n + 2) + (n + 2) + n + n$. Carlos thinks that he can represent the total number of tiles by $4(n + 1)$. What do you think about Carlos and Maria's representations? How do you think they would explain their thinking? Can you show if their expressions are equivalent or not?

Solving Equations

Some algebraic equations can be solved using simple mental math. For example, if you were given the problem $x + 2 = 5$, then you would be able to guess that the answer is 3 without really doing any math on paper or with a calculator. You are doing algebra when you solve a problem like this, but you may not realize it. You found the value of the variable that made the equation true.

Solve each algebraic equation using mental math.

1. $m + 6 = 10$

2. $12 - x = 9$

3. $3x = 6$

4. $4t + 1 = 25$

5. $2x + 2x = 12$

6. $10x - 1 = 99$

Inequalities I

The solution to a linear inequality in two variables is a half plane. Test the given points below to see if they make the inequality true or false.

Inequality: $y > 3x - 5$

Test Points:

1. (0, 0)
2. (4, 1)
3. (3, 2)
4. (–3, –2)
5. (3, –4)
6. (–2, 2)
7. (0, –5)
8. Graph the inequality and plot the test points on the same coordinate plane.
9. What do you notice about the points that make the inequality true?
10. What do you notice about the points that make the inequality false?
11. What real-world applications involve inequalities?

Dividing by Fractions

Write a word problem that could result in the number sentence below. Then explain how you know your word problem fits the sentence.

$$5 \div \frac{2}{3} = 7R\frac{1}{2}$$

Inequalities II

An inequality is a way of writing an equation that does not have an equal sign. The symbol $<$ means "less than." The symbol $>$ means "greater than." For example, $6 < 9$ means that 6 is less than 9. Like equalities, inequalities must be true. The symbol $\leq$ means a number is "less than or equal to." The symbol $\geq$ means "greater than or equal to." For example, you must be 18 or older to vote. If the voting age is v, then the inequality statement for voting age would be $v \geq 18$.

Write an inequality statement for each of the following.

1. You must be more than 4 feet tall to go on this ride.

2. You have to be 65 or older to collect Social Security.

3. The speed limit on the highway is between 45 and 65 mph.

An Unusual Area

Vachon is helping his uncle tile a patio that has an unusual shape. He has drawn a sketch on grid paper to represent the area that they need to cover. Each unit on the grid paper represents 2 feet. Find the area of the figure, and explain your strategy.

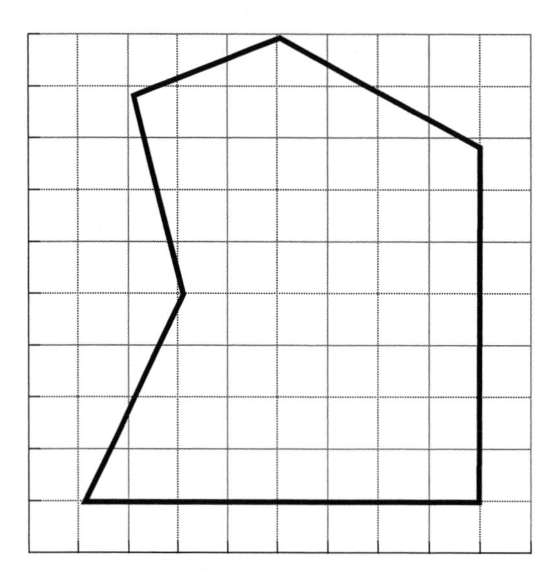

NAME:

GEOMETRY
CCSS 6.G.1

Areas of Triangles

Give the base, the height, and the area of each triangle pictured below. Are the triangles congruent? Explain your thinking.

1.

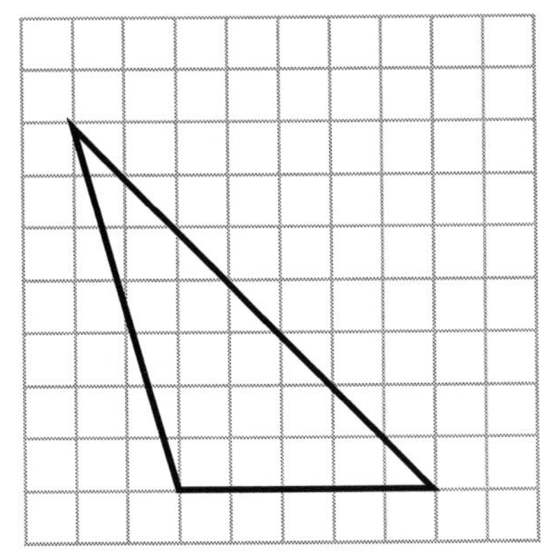

2.

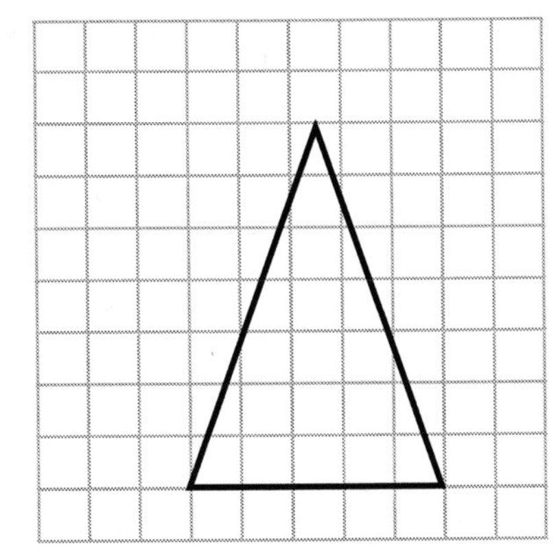

3.

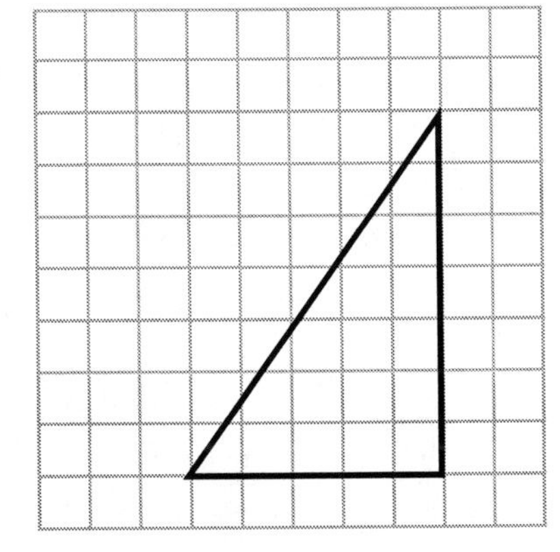

NAME:

GEOMETRY
CCSS 6.G.1

Which Is More?

With a partner, discuss your answers to the following questions and then write them in the spaces below.

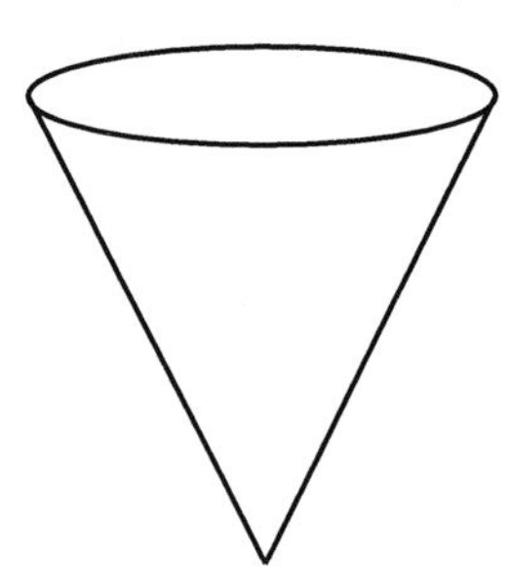

Which paper cup do you think holds the most water? Circle the one that you chose.

What do you think determines the volume of the paper cup? In other words, what determines how much water the cup can hold?

NAME:

GEOMETRY
CCSS 6.G.4

Can Can

What are the different shapes that make up the surface area of the soup can below? Draw lines and write labels to show each shape. Remember, some of the shapes can't be seen in this picture. You need to imagine what the can looks like from different perspectives.

Fold It Up

Here is a net of a rectangular prism. A net is a two-dimensional representation of a three-dimensional object. If you fold up this net, you will see that it makes a rectangular prism.

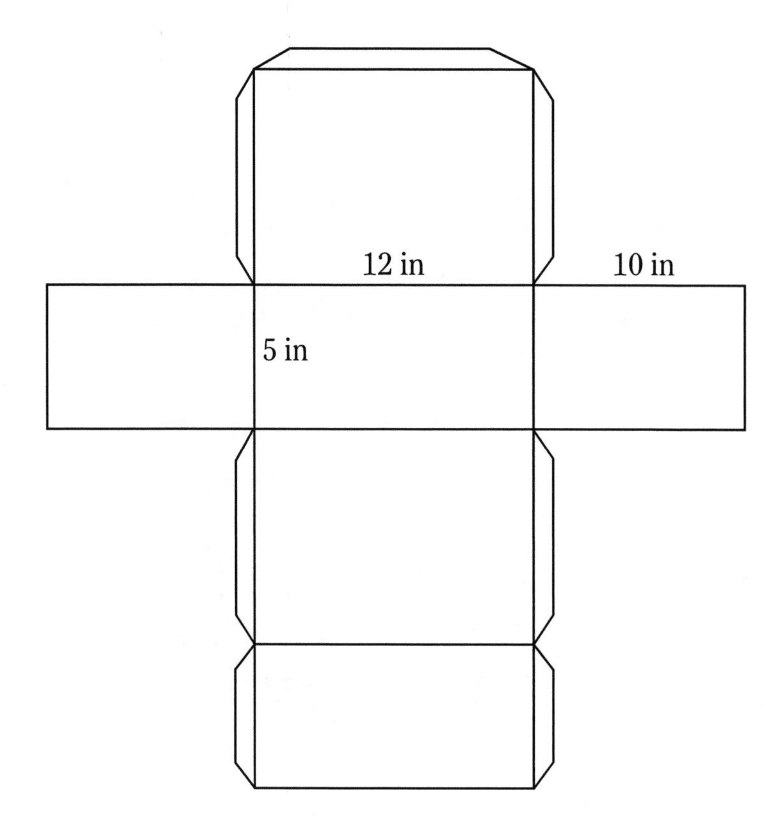

Find the surface area for the rectangular prism using the net above. Show your work and write your answer below.

NAME:

GEOMETRY
CCSS 6.G.4

Take It Apart

What are the shapes that make up the outside surfaces of these objects? Are any of the faces the same? List the shapes below. Circle the shapes that you find in more than one object.

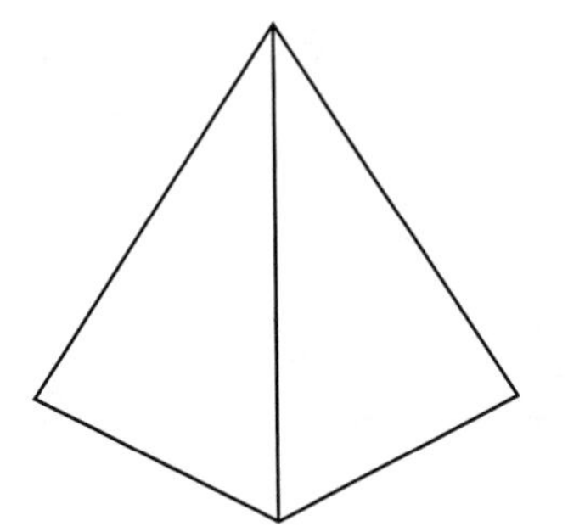 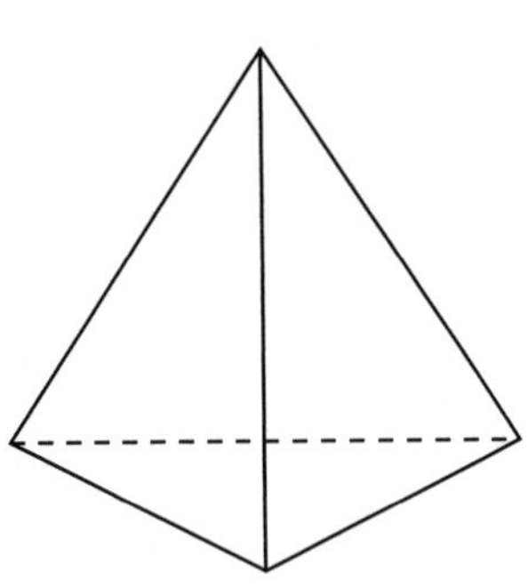 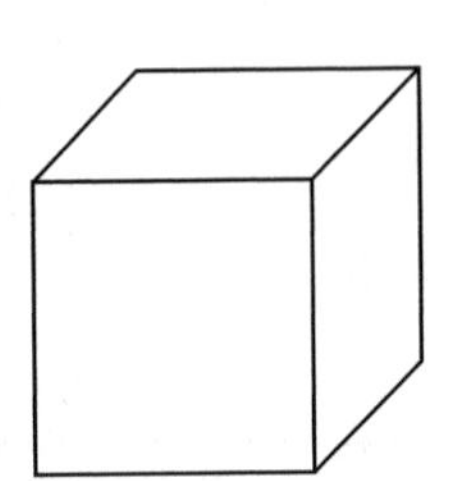 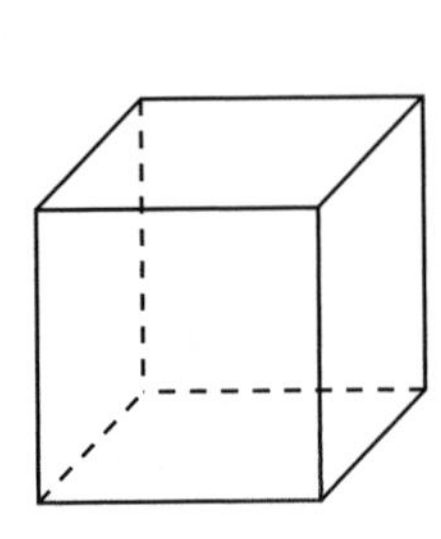

NAME:

GEOMETRY

CCSS 6.G.4

Flatten It Out

What is the surface area of the box whose net is shown below? Show your work and write your answer below.

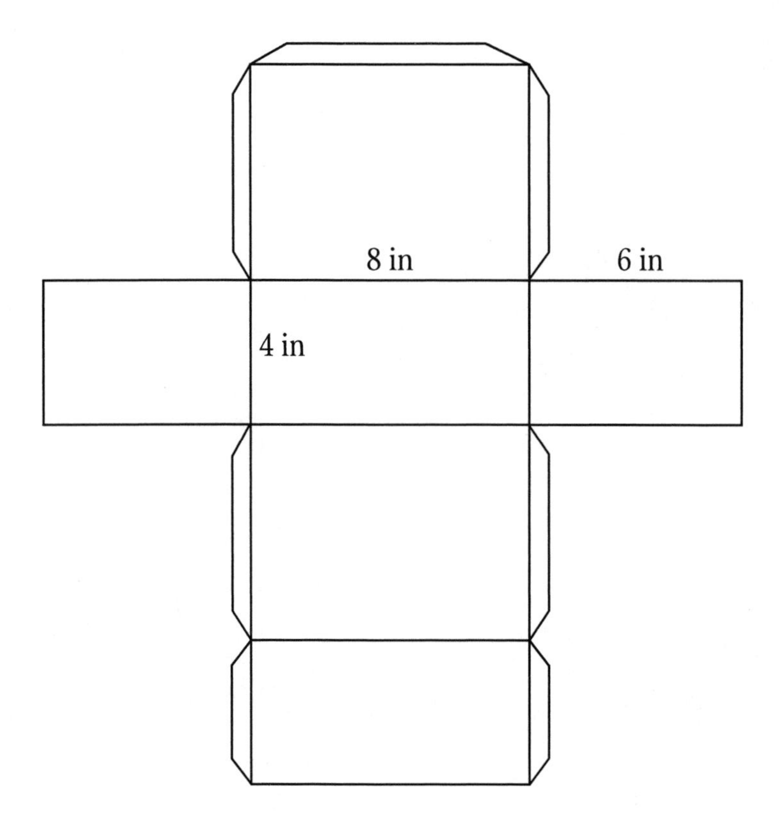

NAME:

STATISTICS AND PROBABILITY

CCSS 6.SP.1

Learn a Language

Read the scenario and answer the questions.

> Administrators at a large high school are trying to figure out which foreign languages to offer. How could they determine which three languages to offer? What information would you want to gather in order to decide which languages to offer?

Write your thoughts and questions below.

At Home with Data

The students in a sixth grade class were asked how many people live in their home including themselves. These are their answers:

2, 4, 4, 5, 2, 3, 4, 3, 4, 7, 3, 4, 2, 5, 4, 3, 6, 4, 3, 5, 4, 4

How can you organize this data so that you can make sense of it? Below, write how you would organize the data.

Organize It!

After you have collected data, you need to organize it. Once you have organized the data, then you can make sense of it. You may want to know which data item occurs the most or which data item occurs the least. One way to organize data is in a frequency table.

Laurel threw a number cube 25 times. Organize her data in the frequency table below.

5, 4, 5, 4, 3, 6, 1, 2, 4, 2, 3, 1, 6, 5, 3, 1, 2, 4, 5, 6, 2, 2, 4, 3, 5

Number	Tally	Frequency
1		
2		
3		
4		
5		
6		

What's the Score?

The histogram below shows the distribution of math test scores in Mrs. Jordan's class. Fill in the frequency table below using the data from the histogram.

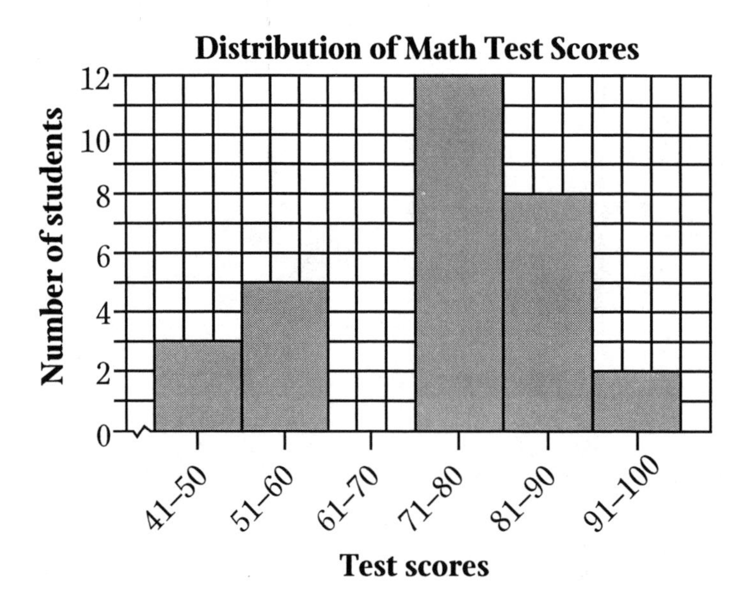

Test scores	Frequency
41–50	
51–60	
61–70	
71–80	
81–90	
91–100	

Predicting with Pennies

Imagine you wanted to know how many heads you would get if you tossed a penny 700 times. You probably would not want to toss the penny 700 times because it would take too long! If you toss a penny just 10 times and count the number of heads, you can predict how many heads you would get if you actually did toss the penny 700 times.

Toss a penny 10 times and record your results. Write a ratio of the number of heads you got to the total number of tosses.

How many heads would you predict you would get with 700 tosses? How did you come up with your answer? Show your work and write your answer below.

Quiz Scores

Parker wants to better understand his quiz score data from last semester. His scores are listed below.

71, 84, 89, 92, 54, 77, 91, 57, 81, 77, 73, 63, 77

What is the mean?

What is the mode?

What is the median?

Which measure of central tendency do you think Parker would like to use to represent his work? Explain why.

Do you think that Parker's teacher would agree? Why or why not?

Answer Key

Ratios and Proportional Relationships

Jumping Jellies!, p. 1

1. 25
2. 25
3. There are the same number of each.
4. 50
5. 200
6. There are 50 pineapple jelly beans and 200 total. 1/4 of the jelly beans are pineapple.

Solving Proportions, p. 2

1. 2
2. 2.5
3. 10 2/3
4. 6

Money, Money, Money, p. 3

Number of hours	Amount earned
1	$8
2	$16
3	$24
4	$32
5	$40
6	$48
7	$56
8	$64

Birthday Roses, p. 4

Students could scale to 60 roses, which yields $75 and $69 respectively, or find the price per rose, which yields $1.25 and $1.15 respectively. Thus, 20 roses for $23.00 is the better buy per rose.

Paolo's Pizza Pricing, p. 5

The 18-inch pizza provides the best value. 9-inch pizza: 63.6 square inches/10.5 = 6.06 square inches per $1.00; 12-inch pizza: 113.1 square inches/15 = 7.54 square inches per $1.00; 18-inch pizza: 254.5 square inches/19 = 13.39 square inches per $1.00

Stamping Around, p. 6

Divide $9 by 20 to get $0.45. In 2012, one stamp cost $0.45.

Faster Than a Speeding Bullet, p. 7

Distance equals rate times time, so $240 = 4r$. In this example, $r = 60$, so Anna's mother drove 60 mph.

Let It Snow, p. 8

$275 \div 25 = 10$; therefore, Estaban has shoveled 10 walkways.

Square Pizza, p. 9

There will be no remainder. The friends predict that they will eat 110% of the pizza (10% + 50% + 35% + 15% = 110%), so there won't be enough.

Playing with Proportions, p. 10

16/20 simplifies to 4/5. 80% = 80/100, which also simplifies to 4/5. Therefore, both fractions are equivalent, and 16/20 is 80%.

How Sweet It Is, p. 11

To compare numbers given in various forms, it is often easiest to put them into the same form. In this example, rewriting all of the numbers in decimal form is most efficient. 30% = 0.3, 23% = 0.23, 1/5 = 0.2, and 2/9 = $0.\overline{2}$. Therefore, the numbers in order from least to greatest are 0.18, 1/5, 2/9, 23%, 0.25, and 30%.

Oil and Vinegar, p. 12

You can set up and solve a proportion: $3/2 = 12/x$. $3x = 24$, and $x = 8$. You will need 8 tablespoons of vinegar. You can also encourage students to come up with the denominator of a fraction that is equivalent to 3/2 which has 12 in the numerator.

Find the Distance, p. 13

Since there are 3 feet in a yard, divide 15 by 3 to get 5. Or, you could set up and solve a proportion: 3 feet/1 yard = 15 feet/x yards = $3x$ and then $x = 5$. Students may also use repeated addition or may share another method. Support all ways.

The Number System

Ribbon and Bows, p. 14

Veronica can make 9 bows, with $\frac{1}{3}$ yard remaining.

Baking Blueberry Pies, p. 15

45 pies (or 44, depending on rounding)

Super Sub Sandwich, p. 16

1. 25 students would get a portion with $\frac{1}{4}$ remaining.
2. 17 students would get a portion.
3. Students should use drawings of the sub divided into appropriate length portions.

Fun with Fractions, p. 17

10; student drawings will vary.

Using 10 × 10 Grids, p. 18

$0.3 \times 0.4 = 0.12$; $\frac{3}{10} \times \frac{4}{10} = \frac{12}{100}$; 30% of 40% is 12%

What Day of the Week?, p. 19

Anatole was born on a Saturday. 759/7 = 108 R3

Greatest Common Factor, p. 20

The GCF is 12. Find the prime factorization of each number and compare. Then take the product of the common factors.

Library Day, p. 21

They will all be there again in 300 days; this number represents the least common multiple (LCM).

Cicada Cycles, p. 22

1. 221 years
2. 336 years
3. The answer represents the LCM.

Counting Cookies, p. 23

1. 10 cookies: yes; 12 bags of 10 cookies
2. 5 cookies: yes; 24 bags of 5 cookies
3. 9 cookies: no
4. 3 cookies: yes; 40 bags of 3 cookies
5. 20 cookies: yes; 6 bags of 20 cookies
6. 7 cookies: no

Factors and Multiples, p. 24

Factors that 12 and 18 have in common: 1, 2, 3, 6
Multiples that 6 and 8 have in common: 24 and 48

Forward and Back, p. 25

Answers will vary. However, encourage students to recognize that whether they walk 7 paces forward or backward, they are covering the same distance.

Balloon Flight, p. 26

5,280 feet

Scuba Duba Doo, p. 27

The two were 2,430 feet apart. Student drawings should show Tanya on a mountain and Olga under water. Ground level would be 0.

Expressions and Equations

As Easy As ABC, p. 28

Students will come up with a variety of answers. After giving them time to work, have students write the expressions they came up with and their answers on the board. Have students check the accuracy of the work that is put up on the board and correct any errors they find.

Powers and Exponents, p. 29

1. 9^4
2. h^7
3. m^2n^5
4. 10^4
5. 625
6. 64

A Number Puzzle, p. 30

Students should get 6 each time. $\{[n + (n + 1) + 11]/2\} - n = 6$

Do You Speak Math?, p. 31

Math	Words
$y - 2$	y minus 2
$7 + z$	7 plus z
$2p$	2 times p or 2 multiplied by p
$3m$	the product of m and 3
$c/5$	the quotient of c and 5
$x/4$	x divided by 4
$t + 9$	t plus 9, or t increased by 9, or t and 9, or the sum of t and 9
$8 + v$	8 more than v
$6 + b$	6 increased by b
$10 - r$	10 decreased by r, or 10 minus r, or r less than 10

Writing Equations, p. 32

1. 62 (for example) $= b + 13$
2. $C = 3x + 2y$
3. $T = r/2 - 500$
4. $T = \$10.50x$

Collette's Chocolates, p. 33

1. Drawings will vary, but should reflect the pattern.
2. 15 orange creams; $(R - 1)(C - 1)$ = number of orange creams for R rows and C columns of French mints

Evaluating Simple Expressions I, p. 34

1. -3
2. $-5\frac{1}{2}$
3. 128
4. 4.75
5. 0
6. 72
7. 1
8. 16
9. 2
10. 50

Evaluating Simple Expressions II, p. 35

1. 24
2. 10
3. 413
4. 24
5. 12.5
6. -90
7. 3/4
8. 0
9. 0
10. -40

Using Properties, p. 36

1. $4 + (91 + 9)$
2. $(3 \cdot 10) + (3 \cdot 2)$
3. $2 \cdot 84$

Border Tiles, p. 37

Maria may be counting all the tiles at the top and bottom of the border $(n + 2)$ and then adding the remaining tiles on each side (n) for her expression. Carlos may be thinking that the border can be separated into 4 equal sets of $(n + 1)$ sections.

Solving Equations, p. 38

1. $m = 4$
2. $x = 3$
3. $x = 2$
4. $t = 6$
5. $x = 3$
6. $x = 10$

Inequalities I, p. 39

1. (0, 0) true
2. (4, 1) false
3. (3, 2) false
4. (–3, –2) true
5. (3, –4) false
6. (–2, 2) true
7. (0, –5) false
8. See graph below.

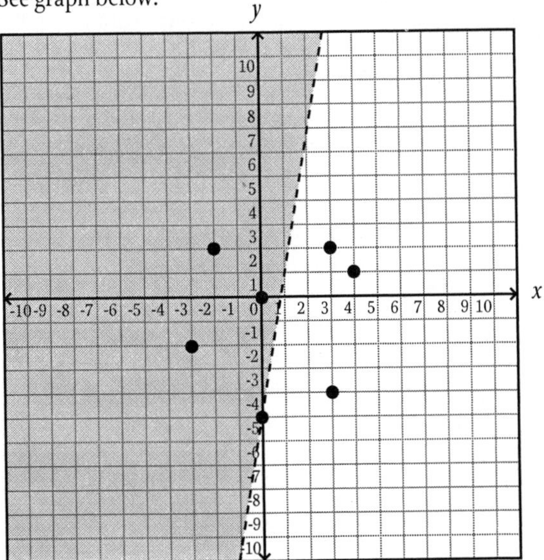

9. The points that make the inequality true lie above the line.
10. The points that make the inequality false lie below or on the line.
11. Answers will vary.

Dividing by Fractions, p. 40

Answers will vary. Sample answer: Sam needs $\frac{2}{3}$ yard of fabric to make an apron. If she has 5 yards of fabric, how many aprons can she make? She can make 7 aprons with $\frac{1}{2}$ of a yard remaining.

Inequalities II, p. 41

1. $h > 4$ feet
2. $a > 65$
3. $45 < S < 65$

Geometry

An Unusual Area, p. 42

The area is approximately 220 square feet. The area can be found by separating the figure into rectangles and triangles.

Areas of Triangles, p. 43

For each triangle, $b = 5$, $h = 7$, and $A = 35$. They are not necessarily congruent; however, their areas are equal.

Which Is More?, p. 44

Students should recognize that both the size of the base and the height contribute to how much water the glass can hold.

Can Can, p. 45

Students should notice that the surface area of a can (cylinder) is made up of one rectangle and two circles. You may want to introduce the word *net,* explaining that it is a two-dimensional figure that can be folded up to make a three-dimensional figure.

Fold It Up, p. 46

There are two rectangles that are 5 inches by 10 inches. There are two rectangles that are 10 inches by 12 inches. There are two rectangles that are 12 inches by 5 inches. So the areas of the six rectangles are: $(5 \times 10) + (5 \times 10) + (10 \times 12) + (10 \times 12) + (12 \times 5) + (12 \times 5)$. The total surface area is $50 + 50 + 120 + 120 + 60 + 60$, which equals 460 square inches.

Take It Apart, p. 47

Help students recognize which faces on each shape are the same. For example, a cube has 6 faces that are all the same. A triangular pyramid's three side faces are the same. A triangular prism's side faces are the same and the two bases are the same.

Flatten It Out, p. 48

$2[(4 \times 6) + (6 \times 8) + (8 \times 4)] = 208$ sq. inches

Statistics and Probability

Learn a Language, p. 49

Students should recognize that they could take a survey of the students in the school and ask them what language they would like to take. This could be accomplished by providing the students with a list of possible choices, such as French, Spanish, Chinese, Latin, and Italian. Look for students to understand that if they take a survey, the survey results will let them know which languages are the most popular and therefore, will let the school know which three languages to offer. Students should see that there is a purpose for collecting data.

At Home with Data, p. 50

Students may choose to put the numbers in order from least to greatest, or they may decide to put all of each number together (that is, all the 2s together, all the 3s together, etc.). Help them recognize that if their data is organized, they can make observations about it more easily. One such observation might be that families of 6 or 7 are not that common or that a family of 4 is the most common.

2, 2, 2

3, 3, 3, 3, 3

4, 4, 4, 4, 4, 4, 4, 4

5, 5, 5

6

7

Organize It!, p. 51

Number	Tally	Frequency
1	III	3
2	IIIII	5
3	IIII	4
4	IIIII	5
5	IIIII	5
6	III	3

What's the Score?, p. 52

Test scores	Frequency
41–50	3
51–60	5
61–70	0
71–80	12
81–90	8
91–100	2

Predicting with Pennies, p. 53

If you get 4 heads out of 10 tosses, your ratio of heads to total number of tosses is 4/10. Students may be able to figure out how many heads they would expect to get in 700 tosses if the ratio remained the same by finding an equivalent fraction to 4/10 whose denominator is 700. Or, you might encourage students to set up and solve a proportion: $4/10 = H/700$ where H represents the number of heads. Because 10 goes into 700 evenly, students do not need to use cross-multiplication to find the value of H. They can use equivalent fractions. $10 \times 70 = 700$, so $4 \times 70 = 280$, which would be the number of heads you would expect in 700 tosses. If students know how and are comfortable solving proportions, they can cross-multiply to get $10H = 2{,}800$. Then, when they divide both sides by 10, they get $H = 280$.

Quiz Scores, p. 54

Mean: 76; mode: 77; median: 77; answers will vary.